Brewster Buffalo

T0324131

First Brewster

The XSBA-1 was Brewster's first design. It was a light naval scout bomber. (NASA -Langley)

Introduction

The Brewster XF2A-1 naval fighter made its first flight in December 1937 because the U.S. Navy Bureau of Aeronautics had reconsidered the idea that monoplane aircraft were unsuitable for aircraft carriers because of their high landing speed. The Brewster F2A-1 was to become the very first U.S. monoplane fighter for use on aircraft carriers. Unfortunately it was already more or less obsolete and retired from first line duty at the U.S. Navy when the Pacific war broke out. It had a very unlucky career when it was used by the U.S. Marines in combat against the Japanese Mitsubishi A6M 'Zero' fighter. For this reason some aviation historians consider the Brewster fighter as a failure, but in fact this was undeserved. In the hands of well-trained Finnish air force pilots this stubby little fighter was a great success. Also in the war between the Netherlands East Indies with Japan, the B-339 as it was called by the Dutch was the best fighter they had. It definitely was inferior to the Japanese Zero fighter, but it could absorb a lot of enemy fire before it was downed!

Brewster Aeronautical Corporation:

Brewster & Co was a U.S. based company manufacturing originally carriages. Later they changed their production to automobile bodies. In 1932 James Work purchased the aircraft division of Brewster & Co. He changed the name into Brewster Aeronautical Corporation. The company was based in multi-story urban factory buildings in Long Island City, New York, across the East River from Manhattan. Being an aeronautical engineer, Work started building business

Production

The XSBA-1 was produced in small numbers as the SBN-1 by the Naval Aircraft Factory. (U.S. Navy)

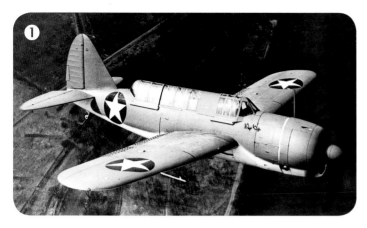

with contract work for other aircraft manufacturers such as making wing panels for the PBY Catalina, but already in 1934 Brewster's design team, headed by Dayton T. Brown, started with the design and construction of their first aircraft. It was a single engine two-seat scout bomber, that received the naval aircraft designation SBA-1. The prototype XSBA-1, Bureau Number 9726, made its first flight on 15 April 1936. Although a small order for 30 machines was placed, these were eventually built by the Naval Aircraft Factory as the SBN-1.

The next type Brown's team designed was a single seat monoplane fighter. It was clearly based on the SBA-1 having the same type of engine. It was built as an answer to a U.S. Navy requirement for a new shipboard fighter. First flight of the new fighter, designated as the XF2A-1, took place in December 1937. It was ordered by the U.S. Navy as the first monoplane type capable to operate from aircraft carriers and it is the subject of this book.

❶-Buccaneer

The Brewster SB2A-2 Buccaneer was produced as a naval attack aircraft.
(Mark Nankivil collection)

❸-Licence Corsair

Brewster licence manufactured the Vought F4U-1 Corsair early model with the 'bird-cage' canopy as the F3A-1.
(U.S. Navy)

❷-400 mph fighter

Brewster's Model 33a was a project for a twin-boom shipboard fighter with a pusher propeller. It was never built.

❹-Attack bomber

The last Brewster aircraft was the XA-32 attack bomber for the U.S.A.A.C. Only two prototypes were build. The photo shows no. 42-13569, the second machine. (U.S. Air Force)

Almost Bermuda!

The SB2A-1 was exported to the United Kingdom as the Bermuda I. On this picture we see British export machines, both in U.S. and R.A.F. markings.
(Mark Nankivil collection)

Left:
*The XF2A-1 prototype in
its original form with small
vertical tail and a 1000 hp
engine*
(©Srecko Bradic)

Right:
*This is the original prototype
at a later stage with enlarged
vertical tail, changed cockpit
and a 1200 hp engine designated as XF2A-2.*
(©Srecko Bradic)

Wind tunnel

*The XF2A-1 being tested in the full-scale
NACA wind tunnel at Langley. This photo
was taken on 5 February 1938.*
(NASA - Langley)

The next Brewster design was a more powerful scout bomber with a Navy type designation of SB2A. The prototype made its first flight on 17 June 1941. It was built on a relatively small scale for the U.S. Navy, where it became known as the Buccaneer. A few hundred were ordered for export to Great Britain as the Bermuda I and to the Netherlands East Indies (where they eventually never arrived). Brewster also designed in 1941 for the U.S Navy a '400+ mph' single seat fighter as the Model 33A. It had a twin tail boom with a pusher propeller and with the cannon armament concentrated in the nose. It was never built!

Under a license contract the Brewster Aeronautical Corporation also built, during the war years, 735 Vought F4U Corsair fighters as the Brewster F3A-1.
The last Brewster design was for a single seat attack bomber for the U.S.A.A.F. powered by a single Pratt & Whitney R-2800-37 Double Wasp of 2100 hp. It received the type designation A-32. Two XA-32 prototypes were built and flown in 1944 before it was cancelled. The XA-32 was the last type built by Brewster. After reporting a large loss, the management decided to shut down the company, and on 5 April 1946 the Brewster Aeronautical Corporation was dissolved by its shareholders.

Prototype in flight

Beautiful in-flight shot of the XF2A-1 prototype.

(Mark Nankivil collection)

The Brewster Buffalo - Early development and test flying:

In 1936 the last biplane fighter for use from aircraft carriers entered service at the U.S. Navy. It was the Grumman F3F and it remained in service until 1941. In spite of earlier objections from the U.S. Navy staff it was evident the biplane era was definitely over. The greatest objection against a monoplane fighter for shipboard use was its high landing speed, but it was more than evident that the biplane fighter was just too slow to be on even terms with enemy fighter aircraft. The Navy set up requirements for a new shipboard fighter and both Grumman and Brewster submitted design proposals. Brewster's new fighter, designated XF2A-1, featured a short barrel-shaped fuselage with straight wings and a main undercarriage that retracted into the fuselage. The Brewster design was constructed conforming to the latest techniques at that time. It was an all-metal construction with a monocoque fuselage.

As a naval fighter it was fitted with an arrester hook for deck landings but the wings could not be folded. The Grumman proposal, designated XF4F-1, was a biplane development of the F3F,

but Grumman quickly realized that a biplane could not compete with the Brewster monoplane, and submitted a new monoplane proposal; the XF4F-2.

Improved 1

The improved prototype XF2A-2 with a more powerful engine and enlarged tail fin.

(Mark Nankivil collection)

Improved 2

The XF2A-2 in flight with the radio mast still on the port side. On the production models F2A-2 it was moved to the starboard side of the fuselage. (Mark Nankivil collection)

Production
The following F2A's and export models were manufactured by Brewster:

Designation	For	No.	BuNo/Serial	Delivered
XF2A-1/2	U.S.Navy	1	0451	Jan 1938
F2A-1	U.S.Navy	11	1386-1396	Oct-Dec 1939
Model 239	Finland	44	BW-351 to BW-394	Jan-Feb 1940
Model 339B	Belgium	40	(diverted: RAF/FAA)	Apr-Jul 1940
F2A-2	U.S.Navy	43	1397-1439	Sep-Nov 1940
Model 339E	RAF	170	W8131-250, AN168-217	Jan-May 1941
Model 339C/D	Netherlands	72	B-395 to B-3167	Mar-June 1941
F2A-3	U.S.Navy	108	01516-01623	Aug-Dec 1941
Model 339-23	Netherlands	20	(diverted: RAAF/USAAF)	Jan-Apr 1942

Total production: 509 including the prototype.

Notes:
-XF2A-4: prototype of pressurized cockpit; built from F2A-1 BuNo. 1388
-'Delivered' dates are based on varying criteria and are representative only.

VF-3

In flight shot of F2A-1 BuNo.1388 from VF-3.
(Mark Nankivil collection)

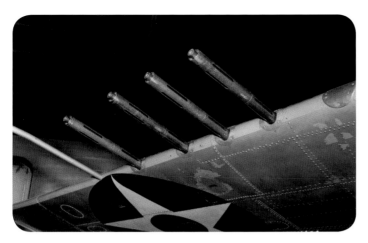

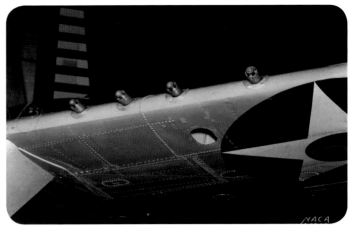

Above: **Armament research**

At NACA-Langley various 8-gun wing armament configurations were tested.
(NASA - Langley)

F2A-1 converted into F2A-2 in all grey colour scheme as used in 1941 by VF-2 Squadron on USS Lexington. Telescope gunsight has been removed!
(©Srecko Bradic)

Wildcat

Grumman's Wildcat was a fighter in the same class as the Brewster F2A. Initially it did not score any better against Japanese aircraft at the outbreak of the war but later versions were substantially improved. This is the XF4F-2 prototype BuNo. 0383 as flown in 1938..

Left: **Join the Navy!**

The U.S. Navy team showed at the New York World Fair of 1939 one of their F2A-1 fighters. (U.S. Navy)

Visit to U.S.A.A.C.

F2A-1's of VF-3 are here visiting an A.A.C. field.
(Mark Nankivil collection)

More VF-2

F2A-2 BuNo.1410, again from VF-2 squadron.
(Mark Nankivil collection)

Nose stand

Not all deck landings were good landings as we can see! . The unlucky pilot on this occasion was Lt. John Smith 'Jimmy' Thach. Aircraft is F2A-1 BuNo.1393 and date and location were 19 March 1940 on the flight deck of USS Saratoga. Thach gained later fame during the Battle of Midway as a Wildcat pilot. (Mark Nankivil collection)

VF-2

Nice in flight shot of a VF-2 F2A-2 BuNo.1406. (Mark Nankivil collection)

Collapse

The retractable undercarriage of the F2A-1 was a weak point and caused problems throughout its whole service career as a naval fighter during deck landings.

(Mark Nankivil collection)

Into U.S. Navy service:

The Brewster F2A-1 was the first monoplane fighter for use on U.S. aircraft carriers. Originally the U.S. Navy placed an order for 54 F2A-1s, but only eleven (BuNos. 1386 to 1396) were actually delivered over the last months of 1939 to U.S. Navy squadron VF-3. The remaining aircraft were supplied to the Finnish Air Force as Model 239, although the BuNos. 1397 to 1440 were already assigned. Officially the F2A-1 never received a name, but the Navy eventually adopted the British name for its export versions, the 'Buffalo'. The F2A-1 was fitted with a 950 hp Wright R-1830-34 Cyclone and a Hamilton Standard variable-pitch 3-bladed propeller. Nine of the F2A-1's were later modified to F2A-2 standards and served with VS-201 on the USS *Long Island*.

When the original XF2A-1 prototype was fitted with a 1200 hp R-1820-40 engine it was re-designated as XF2A-2. Other changes included a Curtiss Electric

Norfolk

CV 7 Air Group at NAS Norfolk 1940 with a single F2A-1 (BuNo.1386). It was most likely used for solo fighter training. (Mark Nankivil collection)

Rudder gone!

F2A-2 BuNo.1422 from VF-2 has lost most of its rudder. Most likely this was caused by a propeller of a neighbouring aircraft with a careless pilot! (Mark Nankivil collection)

propeller and powerplant changes for better performance at higher altitudes. Operational testing of the XF2A-2 started in the summer of 1939 and production deliveries of the F2A-2 started in September 1940. It was delayed by Brewster's concentration on the export order for Belgium. In total 43 F2A-2's were ordered with BuNos. 1397-1439

They served at VF-2 and VF-3 squadrons. Although the F2A's were fitted with inflatable wing 'balloons' in the wings in case an emergency landing had to be made on the water, they lacked any pilot armour and self-sealing fuel tanks until emergency modifications were made in late 1941. With its more powerful engine the F2A-2 could reach a top speed of 550 km/h at 5000 m. Some F2A-2s were fitted with a camera for armed reconnaissance missions as F2A-2P. The F2A-2's saw very limited active service as a fighter during the war, although as trainers some were used until late 1943.

VS-102

This F2A-2 carrying the registration 201-S-13 was one of the eight rebuilt from an F2A-1. When this picture was taken in June 1941 it served at VS-201 Squadron. It was operated from USS Long Island.

(Mark Nankivil collection)

Langley test plane

The Brewster F2A-2 BuNo.1426 was used by NACA at Langley as a test plane for a full-span 'flaperon'. This photo was taken on 9 February 1943 at Langley Field.

(NASA - Langley)

Splinter camouflage

This VF-3 F2A-2 was one of the two that were during August and September 1940 fitted with an experimental splinter camouflage proposed by a U.S. painter named McClelland Barclay. It remained an experiment!

(Mark Nankivil collection)

Flight deck crash

A VF-2 F2A-2 after a less successful deck-landing on USS Lexington. Aircraft BuNo.1406 was flown by the squadron leader. This photo was taken in 1940 and it still shows the old style national marking, but the light all-grey colour scheme.

(Mark Nankivil collection)

Unidentified

An unidentified F2A-2 with test 'flaperons'; most likely the same as BuNo. 1426, at a different date.

(Mark Nankivil collection)

The last Navy operational version was the F2A-3. Originally it was intended as a scout fighter with fuel tanks of 300 litres increased capacity, with two self-sealing 75 litre tanks in the wing leading edge and a 150 litre tank under the pilot's seat. As a consequence the typical ventral windows of the other versions could not be used. Instead, on the F2A-3 the framed window was plated. It must be remarked that also on a number of Finnish Buffalo's the windows were replaced by plating, most likely because of lack of spare window panels! The F2A-3 also carried more ammunition and at least the pilot was protected by armour plating. Unfortunately this increased the aircraft's all-up weight with some 450 kg without being compensated for by more engine power. It not only resulted in a reduced top speed and rate of climb, but it also degraded the turning and manoeuvring capacity. The increased weight also caused more landing gear failures during carrier landings and although the F2A-3 saw some service on the aircraft carriers *Saratoga* and *Lexington* the U.S. Navy had decided before the attack on Pearl Harbor to retire its Buffalo's from operational service as a carrier fighter. They were mostly sent on to the Marine Corps or, just like the F2A-2s , used as advanced solo trainers.

Family man

In October 1940 a number of F2A-2 were ferried from the Brewster plant to Anacostia NAS by U.S. Navy pilots who were ordered to carry civil clothes for 'security reasons'. Apparently we see one of these ferry pilots, most likely Howard Packard, during a transit landing with his family.
(Mark Nankivil collection)

Pearl Harbor

Interesting shot at Pearl Harbour before the Japanese attack. A single F2A-2 can be distinguished in front of the open hangar door.
(Mark Nankivil collection)

Formation dive

A formation dive of a group of F2A-3's from a training unit over Florida in August 1943.

Anonymous

An anonymous F2A-2 in late 1942 colour scheme with removed propeller spinner.

NAS Miami

An F2A-2 carrying a large identification number as used by the Carrier Transitional Training Unit at Naval Air Station Miami. Note modified radio antenna without the characteristic mast, indicating prior use by VF-2.

(Mark Nankivil collection)

No markings!

A peculiar photo of an F2A-2 without U.S. national markings on the fuselage. Number '06' served as a trainer. (Mark Nankivil collection)

BuNo.1412

Action shot of VF-2 F2A-2 BuNo.1412.
(Mark Nankivil collection)

Left: **Guns loaded**

Ordnance man loading the guns of an F2A-3, most likely also Miami 1943.

(Mark Nankivil collection)

NACA

The NACA at Langley also flew with the F2A-3 BuNo.01516.

(NASA - Langley)

The U.S. Marine Corps deployed two F2A-3 squadrons to the Pacific, VMF-112 at Palmyra Atoll, and VMF-221 at Midway Island. F2A-3's from VMF-221 first saw action in early March 1942, when four Brewsters shot down a Kawanishi H8K "Emily" flying boat. More significantly, the F2A-3's provided aerial defence on 4 June 1942 during the Battle of Midway against Japanese fighters and bombers from four carriers. They had some success: Captain W. C. Humberd shot down a Zero during a frontal attack. Other VMF-221 pilots successfully disrupted a dive bomber attack made by Aichi 'Val' bombers. Captain Marion E. Carl (flying an F4F-3) also claimed a Zero as destroyed. Carl would end the war as an ace with 18 ½ aircraft destroyed.

Red tail

F2A-3 BuNo.01553 '2' as used at NAS Miami for solo training; this particular aircraft had previously flown with the Marines at Midway. (Mark Nankivil collection)

Unlucky landing

This VMF-221, no. MF-5, made an unlucky landing on USS Long Island.
(Mark Nankivil collection)

NAS Pensacola

F2A-3 with the code number '21' photographed over NAS Pensacola in August 1943. It was flown by Lt. Cdr Joseph Clifton. The rear part of the canopy has been painted over to reduce heat.

(Mark Nankivil collection)

However, in their battle reports all pilots clearly stated they regarded their Buffalo's as inferior to the Japanese Zero regarding speed, climbing performance, armament and manoeuvrability! It must be stated that the F4F Wildcats from VM-221 definitely did not do much better than the F3A-3. That the Battle of the Midway was not very successful for the F2A-3 is clearly shown in the VMF-221 losses: in total 13 of the 21 F2A-3's had to be written off because they were shot down or damaged beyond repair!

The losses were as follows: (right)

By the second half of 1943 all remaining F2A-3s in use at the various Naval Air Stations in the U.S. as trainers were withdrawn from use and subsequently scrapped. There are no survivors present in the United States.

Stud

'F2A Instructing studs at Miami 1943' is the title of this rare colour shot. It is an F2A-3 which had less framing on the sliding part of the canopy.

(Mark Nankivil collection)

On hoist

F2A-3 being hoisted on board of a transport carrier by a crane for transport to Midway. The aircraft was assigned to U.S. Marine squadron VMF-221.

(Mark Nankivil collection)

Lost airplanes

Squadron Code	Type	BuNo.	Pilot
MF-1	F2A-3	01518	Maj.Parks
MF-19	F2A-3	01520	Cap. Curtin
MF-7	F2A-3	01522	Cap. Hennessy
MF-9	F2A-3	01524	Cap. Merrill
MF-3	F2A-3P	01525	Cap. Alvord
MF-10	F2A-3	01528	2Lt. Benson
MF-4	F2A-3	01537	2Lt. Butler
MF-8	F2A-3P	01541	2Lt. Lindsay
MF-12	F2A-3	01542	2Lt. Lucas
MF-1	F2A-3	01548	2Lt. Madole
MF-18	F2A-3	01559	2Lt. Mahannah
MF-14	F2A-3	01563	2Lt. Sandoval
MF-5	F2A-3	01569	2Lt. Pinkerton

F2A-3 'MF-15' as flown by Capt. Humberd from U.S. Marines VMF-221 during the Battle of Midway in June 1942. Humberd shot down a Zero fighter during a head-on attack.
(©Srecko Bradic)

BuNo. 1410

F2A-2 no.1410 from VF-2 in early colours of Yellow 6th section on board of USS Lexington. (Mark Nankivil collection)

First encounter
Brewster F2A-3 vs Mitsubishi Zero

During the Battle of Midway in June 1942 Captain W.C. Humberd was the first U.S. Brewster pilot scoring a confirmed kill on a Japanese fighter, a Mitsubishi Zero.
This battle scene is dramatically pictured on the front cover by Srecko Bradic as an acryl painting.
The statement of Captain W. C. Humberd, USMC in his battle report, given on June 4th, 1942, is the best description of this battle scene (typo's included!):

New camouflage

An unidentified F2A-3 with the new two colour camouflage scheme and the later style national markings without the red dot. Note absence of spinner, telescope gunsight and the radio mast. (Mark Nankivil collection)

While in the standby division on morning of June 4, 1942, the air raid alarm sounded at 0559. Our division took off at approximately 0605. In our division of six planes, Capt. Kirk Armistead is division leader, 2nd Lt. W. B. Sandoval his wingman, myself section leader of second section with 2nd Lt. W. V. Brooks as wingman, 2nd Lt. C. Kunz 3rd section leader with 2nd Lt. M. E. Mahannah his wingman. We took off immediately after fourth division and started gaining altitude in direction of approaching enemy which was 310 degrees, altitude 12,000 feet given by base radio.

Sight contact was made of enemy formations at approximately 12,000 feet bearing about 30 degrees to port and distance of about 10-15 miles. We continued climbing to 17,000 feet, still keeping the enemy slightly to our port, then when in position of about 3,500 to 4,000 feet above and still to port we made attack, about 30-35 miles bearing 320° from islands.

By time to make attack, my division leaders wingman had dropped back some in which case I was second to attack. I followed the division leader in a high side approach shooting down one (1) bomber in this approach, then coining up for high side approach on other side I again attacked, thinking I might have shot down another bomber in this approach. I came up on other side and started another approach when, about half way through run, I heard a loud noise and turning around I saw a large hole in hood of my plane and also two type 00 navy fighters on me about 200 yards astern, then I immediately pushed over in steep dive in which one (1) followed me. I descended to water level in trying to gain distance on the fighter, the plane staying with me; I stayed at water level with full throttle gaining distance slowly until I decided the distance was great enough to turn on the plane in which case we met head on, I gave a long burst when we were about 300 yards distant and the plane caught on fire and out of control dived in the water. By this time I was approximately 40 miles from first attack and started gaining altitude up to 10,000 feet. My fuel and ammunition were fairly low, about three-fourths exhausted, and I called to see if field was clear for landing, in which case I received an "affirmative". In the meantime, while climbing for altitude, I discovered my hydraulic fluid had been lost and my flaps and landing gear would not lower so I used emergency system and the wheels lowered, then made proper approach to field and landed. After refueling and rearming, I again took off and while I knew my wheels would not retract, I intended going some distance from field to remain for a period when orders to land were given to all fighting planes.

My plane was a F2A-3, Bureau Number 01553, loaded with 1300 rounds of 50 cal. ammunition, one ball, 2 armor piercing. The attack was made at approximately 0625 and I used approximately 400-600 rounds of ammunition; the final landing being about 0745.

The enemy formations were of a Vee consisting of about five to nine planes each, there being about 4 to 5 of such formations in group we attacked. I don't know what formation the fighters used or where they were as the first I knew of their presence was the loud burst in my plane and turning, saw them. The type of bombers seems to correspond to the type 99 Aichi (navy), and the fighters were navy type 00.

After my second approach, I saw about four or five planes going down in flames and only identified one as our own, all this was just a glance on my part. Their fighters seemed to out maneuver us in most all repects except in my case, I out dived the one after me and gained distance at sea-level. Frankly, I think the F2A-3 does not compare with their type 00 fighters whatsoever.

My plane had a number of holes in it, three or four making the left beam tank unusable. Had two large holes in fuselage of what appeared to be 20 mm size. No apparent damage to plane except for left beam tank and hydraulic lines broken.

VF-3

F2A-2 BuNo.1399 with markings of VF-3 naval fighter squadron. (Mark Nankivil collection)

Fuelled up

A U.S. Marines F2A-3 being fuelled up at Ewa Island, Hawaii early 1942.

(Mark Nankivil collection)

Belgian in the U.S.

An unknown photo of the Belgian Buffalo NX95B from the archives of Dade Brothers Inc. at Long Island, NY. This company was responsible for the crating of many export aircraft. This particular aircraft was finally shipped to the UK where it flew with R.A.F. serial BB450. (Cradle of Aviation Museum via Gerard Casius)

The Belgian order:

On 11 December 1939 Brewster was able to proudly announce its first export order. Belgium had placed a contract for the construction and supply of 40 Brewster Model 339Bs. This was basically the F2A-2, fitted with a 1000 hp Wright R-1820-G-105 engine approved for export use. This engine produced some 200 hp less than the Wright R-1820-40 intended for the U.S. Navy F2A-2. All naval equipment (arrester hook, telescopic gunsight and an inflatable rescue dinghy) was removed. The aircraft was further modified with a slightly longer tail cone. They were scheduled for delivery to Belgium in the spring of 1940.

The December 11, 1939 contract between Brewster and the Belgian Ministry for National Defence, Aviation Department, also covered license production. The Belgians were to notify Brewster that they were taking up that option by June 30, 1940, and if so they would have to pay Brewster $100,000 plus $2,800 per aircraft completed. The contract was very clear that this license did not include engines or engine equipment, instruments or armament. However, because of the outbreak of the war in Europe and the occupation of Belgium by German troops in May 1940 no license production was realized. Even the Belgian Model 339B's built by Brewster were never supplied to the Belgian air force, although the order was fully completed by Brewster. They were all test flown at Brewster carrying a civil registration.

Belgian in Germany

This Buffalo wreck was photographed by Richard Bueschel in 1946 at Darmstad, Germany. It is one of the two Belgian Brewster 339's that were delivered at Bordeaux, France, just after the French armistice. There are unconfirmed rumours one was flown in Luftwaffe markings, but this fuselage still bears the original Belgian national markings!
(Daniel Brackx collection)

When the first Model 339B was completed it was packed and shipped to Europe in mid-April. However on 10 May 1940 Belgium was occupied by German forces. After the war a wrecked Model 339B was found near Darmstadt still fitted with its Belgian markings! It must have been one of the two aircraft that finally reached France after the armistice. Six more 339B's were loaded in June 1940 on board of the French aircraft carrier Béarn but they got no further than the Caribbean island of Martinique. This small island had at that time no airfield and the Model 339B's remained there for some time until they were scrapped. All other Belgian 339B's were shipped to the U.K.

Leader

4/LLv 24 leader 1Lt. (or Capt.) Per Sovelius in the cockpit of his Brewster fighter: summer 1941. Per-Erik "Pelle" Sovelius scored a total of 12.5 kills.

(Mark Nankivil collection)

Hans Wind

BW-378 of 4/LLv 24 at Vesivehmaa photographed on the ground in summer 1942 at Vesivehmaa. It was flown by 1Lt. Hans Wind; one of the top-scoring Finnish aces with 75 confirmed kills.

(Kari Stenman collection)

In Finnish service:

In the spring of 1939 the U.S. was approached by Finland for the purchase of new fighters in order to modernise its combat air force. The U. S Navy was prepared to divert from its original order of 54 F2A-1's a total of 44 to Finland in exchange for an order of the improved F2A-2. The final contract was signed on 16 December 1939. For use in Finland, all naval equipment like arrester hook

and the rubber dinghy behind the pilot were removed. With the removal of the arrester hook, the slot for the arrester hook was simply faired in. Also the Navy telescopic gunsights were removed. At Brewster, the Finnish F2A-1s received the export designation Model 239 (the 2nd export proposal for 1939). They were fitted with the export-approved Wright R-1820-G5 engines, suitable to run on low-octane (87%) fuel and giving a maximal take-off power of 950 hp. The

final deal further included the delivery of sufficient spare parts, ten spare engines and twenty Hamilton Standard propellers. The Model 239 was not fitted at Brewster with self-sealing fuel tanks and armour. In Finland armoured backrests for the pilot, metric flight instruments and a Finnish Väisälä T.h.m.40 gunsight were built in. Armament initially consisted of one 12.7 mm and one 7.7 mm machine gun in the top of the fuselage nose (by 1943, both were 12.7 mm) plus an additional two

12.7 mm guns in the wings. The top speed of the Finnish Buffalos, as modified, was 478 km/h at 4,750 m, and their loaded weight was 2,415 kg. Some Finnish Model 239's (BW-357 was one) were fitted with retractable skies and one (BW-370) was fitted as an experiment with non-retractable snow skids.

The Finnish fighters were converted from Navy F2A-1's during January-February 1940. By ship they were transported from New York to Bergen in Norway. From there they were transported further by rail to Sweden for final assembly by SAAB at Trollhättan. Here they were flight tested by Brewster and Finnish pilots and finally ferried to Finland.

Finnish ace

1Lt. Jorma Sarvanto in front of his BW-357; summer 1941. Sarvanto was one of the Finnish aces with a total score of 17.
(Mark Nankivil collection)

Below: **Finnish beauty**

Brewster BW-393 of 4/LLv 24 at Rantasalmi in the summer of 1941.
The identity of the girl is unknown!
(Mark Nankivil collection)

Brewster down!

An unidentified Brewster fighter after a belly landing
(Mark Nankivil collection)

ACE
Top ten ranking Finnish aces and their claims while flying Brewsters (total victories in all types in parentheses) were:

Hans Wind	39	(75)
Eino Juutilainen	34	(94)
Jorma Karhunen	26.5	(31)
Lauri Nissinen	22.5	(32.5)
Eero Kinnunen	19	(22.5)
Nils Katajainen	17.5	(35.5)
Eino Luukkanen	14.5	(56)
Martti Alho	13.5	(15)
Erik Teromaa	13	(19)
Lauri Pekuri	12.5	(18.5)

During the Winter War with the Soviet Union of 1939-1940 only five were ready for combat, but during this period they were not yet used. The Finnish pilots regarded their Brewster fighter as easy to fly. When compared with other fighters in service at that time they were maybe not the fastest fighters, but the long range and flight endurance and relatively trouble-free operation made these planes very popular. The 44 Finnish Model 239's received the serial numbers BW-351 to BW-394. The Model 239 was in service from 1941 to 1945 and some even survived the war. Although less manoeuvrable than the Soviet I-15/I-153 and I-16 they scored many kills in the Continuation War with the Soviet Union. Finnish pilots avoided direct dogfights with the smaller, lighter and more manoeuvrable Soviet fighters, but attacked from high altitude with their superior speed and armament. During the war with the Soviet Union a number of Finnish fighter pilots gained the ace status with impressive scores, most of these gained on the Model 239.

Finnish pilot

MSgt. Sakari Ikonen of 4/LLv 24 seated in his Brewster fighter; summer 1941. Note the armour plate behind the pilot seat!

(Mark Nankivil collection)

Black Lynx

BW387 'Black Nine' of 2.LeLv24 was photographed in May 1942 at Tiiksjärvi. It was flown by Yki Korhonen and carries a lynx figure on the fuselage side as squadron badge. The figure painted on the tail depicts an elk.
(*Frits Gerdessen collection*)

Post war

A small number of B-239's survived the war. We can see here BW373 in post-war markings in the summer of 1946 at Kauhava. (*Frits Gerdessen collection*)

The XF2A-4
By Jim Maas

There are some incorrect details about the aircraft - including the claim that it was converted from the first F2A-3.

In fact, it was one of the F2A-1 aircraft, BuNo. 1388; c/n 3, which were converted to F2A-2 standard by Brewster in November 1940, and then further converted by the Naval Aircraft Factory to XF2A-4 standard in August 1941. This seems to have included an extension to the nose to deal with centre-of-gravity problems. The aircraft was changed to include a pressure cabin and turbo supercharger. The specification directed that the overturn pylon behind the pilot would need to be redesigned to "avoid interference with the pressure bulkhead" - which means that there was some kind of bulkhead behind the pilot's head. The telescope gunsight was to be replaced with a reflector gunsight, and armament reduced to a single wing machine gun. The life raft canister and relief tube were to be deleted.

The specification also stated:
"the portion of the windshield directly forward of the pilot shall be of flat, laminated glass. The remainder of the windshield shall be spherical in shape and shall be made of either laminated glass or a transparent plastic. All transparent sections shall be double with provision for passing heated air between panes. A windshield wiper for the outside flat section of the front windshield shall be provided. Provision shall be made for the application of alcohol to the windshield at the wiper."

The XF2A-4 was flown, but photo's do not seem to exist..

Formation

A formation of Brewsters with BW-352 on the fore-ground flown by MSgt Eero Kinnunen of 2/LLv 24 over Tiiksjärvi in the autumn of 1942.

(Kari Stenman collection)

Kokko

A mechanic on the wing of BW-379 flown by 1Lt. Pekka Kokko, his first name on the nose; summer 1941. Kokko had 13.5 confirmed kills on his name.

(Mark Nankivil collection)

Although most of these pilots also scored earlier victories during the Winter War on other fighter types like the Fokker D.XXI, and during the Continuation War with types like the Bf-109 G, the Brewster produced excellent results with a kill ratio of 26:1 over the Soviet aircraft. The first successes with the Model 239 were recorded on 25 June 1941 when they shot down five Tupolev SB-2 bombers out of a formation of 27. The Model 239's were from Fighter Squadron 24. When flying the Brewster 239 this squadron claimed a total of 459 Soviet aircraft as shot down while losing 15 in combat.

However, when hostilities continued and new Soviet fighters were introduced it was evident the Brewster was getting obsolete and from 1943 on they were replaced by much faster Messerschmitt Bf 109 G's. Most Model 239's remained in service although they were withdrawn from the front lines with the introduction of the

Bf 109 G. The last aerial victory made by a Brewster against the Soviets was on 29 July 1944 over the Karelian area.

When Finland made a separate peace with the Soviet Union they had to fight against Germany, being more or less their former ally. Reportedly, the Finnish air force clashed with Luftwaffe Ju-87's and a Ju-88 during the 'Lapland War'. Brewster pilots Erik Teromaa and Oiva Hietala each claimed a Ju-87 on 3 October 1944. They were the last victories to be made by Brewster pilots in World War 2! By the end of the Second World War only seven Finnish Brewsters were still airworthy. They remained in active duty until the autumn of 1948. The last flights of Finnish Brewsters were made on 14 September 1948 with BW-377 and BW-382. They were scrapped in 1952.

A new Brewster Type 239 'BW-360' of the Finnish air force early 1940 at Trollhättan in Norway in Finnish markings but without any camouflage scheme.
(©Srecko Bradic)

Winter

BW360 on 15 March 1940 at Trollhättan.
(Frits Gerdessen collection)

Museumpiece

The Humu as restored for the Finnish Air Force Museum
(Courtesy Finnish Air Force Museum)

The Valtion Humu

The V.L. Humu was a Finnish fighter aircraft, designed by the State Aircraft Factory (Valtion lentokonetehdas) in 1944. Due to the small numbers (44) of Brewster fighters in service in the Finnish Air Force, the Finns wanted to see if they could design a fighter, based on the Brewster. The aircraft designers Torsti Verkkola, Arvo Ylinen and Martti Vainio were called upon to lead the project that received as name Humu. The Humu was based on the Brewster B-239. The Humu was largely constructed out of wood due to scarcity of metals, but the frame was made from steel and its design followed closely the one of the Brewster. To save time the fuselage of the first Humu was in fact converted from an existing B-239 airframe; no. BW-392. Unfortunately it crashed on its first flight on 5 June 1943 killing its pilot Martti Alho.

In spite of this accident a second Humu was constructed. This aircraft carried the serial HM-671. The first flight took place on August 8, 1944. The HM-671 flew altogether for a total of 19 hours and 50 minutes. The aircraft was 250 kilograms heavier than calculated, its engine too weak and the aircraft was simply not of the standard expected from a fighter aircraft of 1944. Also its performances were in general less than the B-239.

The Finnish Air Force ordered 90 Humus. However, the production was stopped in 1944 when only one single aircraft was fully completed.

Main reason the production of the Humu was cancelled was the fact that the Finnish air force had meanwhile purchased from Germany a number of Bf 109G fighters with much better performances than both B-239 and Humu. The HM-671 has been fully restored and is currently displayed at the Central Finland Aviation museum.

Technical details:

Powerplant:	1 × Shvetsov M-63 9-cyl. air-cooled radial of 1000 hp

Dimensions:

-Length:	8.03 m
-Wingspan:	10.67 m
-Height:	3.66 m
-Wing area:	19.40 m²

Weights:

-Empty:	2050 kg
-Max. takeoff:	2895 kg

Performances:

-Maximum speed:	
	420 km/h at sea level;
	430 km/h at 4500 m
-Service ceiling:	8000 m
-Rate of climb:	13.3 m/sec

Armament:

	2× 12,7 mm LKK/42 machine guns

In the Netherlands East Indies:

After preliminary negotiations to obtain Brewster airframes and engines (including trying to buy the ex-Belgian 'refugee' Brewsters) without much success, the Netherlands Purchasing Commission (NPC) finally succeeded in ordering 72 Brewster Model 339 fighters in October 1940. These fighters were for the Militaire Luchtvaart (ML) of the Koninklijke Nederlands-Indische Leger (KNIL), abbreviated as ML-KNIL. Since engine supply was the bottle neck, the first batch of 24 (registration nos. B-395 to B-3118) were fitted with a Wright Cyclone R-1820-G-105A of 1000 hp (Model 339C version). The remaining machines, designated Model 339D (registration nos. B-3119 to B-3166) were fitted with the 1200 hp G-205A from Brewster's stock. These aircraft were shipped without any guns, armour or self-sealing fuel tanks; these critical items would be sent separately and added in the East Indies. It is not widely known that there were plans to station some Brewster 339D's in Suriname (a Dutch colony on the northwest coast of South America) to defend it against possible enemy occupation because of its bauxite mines. Since Suriname was occupied in November 1941 by U.S. forces this did not happen, and the last five Model 339D's were eventually used as instructional airframes in the U.S. In February 1941 an additional 20 Model 339-23's were ordered. They were similar to the F2A-3 without the naval equipment, but with the longer fuselage and increased fuel capacity. Unfortunately they were also fitted with the same 1000 hp Cyclone R- 1820-G5 engines used on the (much lighter) Finnish Model 239's. These aircraft were intended as trainers so 339C/D's could be released to operational units. The ordered Model 339-23's were never delivered to the ML-KNIL. All machines (registration nos. B3-167 to B3-186) were diverted to Australia, along with the first 339D, no. B-3119, and handed over to the U.S.A.A.F. stationed there.

B-396 assembly

Final assembly at Brewster by mating wings and fuselage of the first ML-KNIL Model 339C

(Gerard Casius collection)

When Japan attacked the Netherlands East Indies on 8 December 1941 not all fighter groups operating the Brewster 339C's and D's were ready. Pilot training was not yet fully completed and most pilots did not have many hours on this type. Because of earlier agreements with the British, a squadron of Dutch Brewsters joined the R.A.F.'s Buffalos at Singapore but with the deteriorating situation they were withdrawn back to the south.. Since the Dutch Brewsters only had underwing markings additional orange triangles were painted on the upper wings of the Dutch machines while in Malaya; once they returned to the East Indies these were removed. Together with the Curtiss Model 75 Hawk and the Curtiss CW-21B, the Brewsters had to defend the very large territory of the Netherlands East Indies against invading Japanese forces. En-counters with Japanese planes followed but given the vast area that had to be defended, the ML-KNIL aircraft were spread too thinly to be really effective and over the months to follow the total number of fighters slowly diminished.

Dutch formation in the U.S.A.

Formation of three ML-KNIL Model 339D's flying over New Jersey (Spring 1941). (Gerard Casius collection)

B-3119

ML-KNIL Brewster Model 339D carrying both military and civil registrations B-3119 and NX341B at Newark, New Jersey, most likely during an open day. Note Brewster employees! (Mark Nankivil collection)

ML-KNIL Brewster Type 339D 'B-3119' as test flown in the United States with civil registration NX341B. It was damaged during a landing accident and when it was repaired it finally found its way to Australia. (©Srecko Bradic)

Against the Japanese Army Ki-43 'Oscar' the Brewster was more or less on even terms. The Ki-43 was of course more manoeuvrable with a higher climbing speed, but the Brewster had a sturdier construction and a comparatively heavy armament of four machine guns. Beginning in late December, however, the appearance of the Japanese Navy's cannon-armed Mitsubishi Zero changed the balance. The Zero could out-perform the Brewster on all points except strength and diving and one by one the Dutch fighters were lost. The ML-KNIL did not have any radar equipment operational and that meant they were definitely on second-best terms during air attacks. The Brewster was also called upon to serve as a dive bomber, fitted with two bombs under the wings.

B-396

ML-KNIL Buffalo B-396 photographed at the Brewster plant. Note gun camera on the right hand side of the fuselage and the installed bomb racks under the wings. The gun camera was rarely used in practice. (Gerard Casius collection)

Dutch instrument panel

Instrument panel of a Dutch Model 339 with metric instruments and Dutch text. The panel in the centre is a short check list. The text reads: 'Controleer dit voor start en landing' (Check this before start and landing). (Gerard Casius collection)

Dutch hero

Lt. August ('Guus') Deibel of 2-VLG-V with his Buffalo (serial B-3110) at RAF Kallang, early 1942. Deibel survived the war, but was later killed in a Gloster Meteor.

(Gerard Casius collection)

Below:
Jaap van Helsingen

'Captain J. van Helsdingen, in the uniform of a Lieuten-ant-pilot.
(courtesy KNVvL, Afd. Luchtvaartkennis)

Jaap van Helsdingen
Netherlands East Indies hero!

Jacob Pieter van Helsdingen ('Jaap') was born on 7 March 1907 at Soerabaja, Netherlands East Indies. He joined the ML-KNIL already before the outbreak of the war as a pilot. He was a Curtiss P-6 biplane fighter pilot. On 27 February 1936 he collided in flight with another P-6. Van Helsdingen survived when he used his parachute, but the pilot of the other P-6 was tragically killed.

At the outbreak of the war with Japan he was a commanding officer of a group of Brewster pilots. On 23 and 24 January 1942 he led his pilots during a dive bombing attack on enemy vessels. This attack was highly successful and inflicted quite a lot of unexpected damage on Japanese side. Unfortunately the tide would turn and with fewer Brewsters every time they were under attack it was a quite hopeless task to command his pilots. In spite of this he showed great courage and great leadership. On 7 March 1942, one day before the capitulation of the KNIL, van Helsdingen received the order to give KNIL infantry troops air support. Unfortunately only four Buffalo fighters were airworthy at that moment. For this mission Van Helsdingen

was originally given orders not to be active as a pilot as CO, but he ignored this and together with three other volunteers, Lt. Guus Deibel, Sgt. Gerard Bruggink, and Vaandrig (2nd Lt.) Jan Frederik Scheffer, they took off to give ground support. The story goes that van Helsdingen selected the worst Brewster they had for himself which was described as being 'nothing more than a wreck with wings'.

After take off the were soon discovered by a group of Japanese fighters. Deibel had to make an emergency landing after being hit by enemy fire. Bruggink and Scheffer managed to escape in the clouds and returned to their base. Jaap van Helsdingen was last seen going down into the jungle. His plane and body were never recovered.

After his death van Helsdingen was decorated with the military 'Willems Orde'. A military barracks complex in Hilversum, the Netherlands was named the 'Van Helsdingen Kazerne' after this Dutch hero.

A low quality, but very interesting movie-still image showing Dutch and Commonwealth pilots in discussion at the tail of an ML-KNIL Model 339C. It shows the Anglo-Dutch cooperation at Singapore but most interesting are the extra orange triangles on top of the wings without the usual black lines. They were applied to prevent confusion with enemy aircraft. (Scott Hochstein collection)

Between December 8th and March 8th, 17 ML-KNIL pilots were killed, and 30 aircraft shot down; 15 were destroyed on the ground, and several were lost to misadventure. Dutch pilots claimed 55 enemy aircraft destroyed but this was almost certainly optimistic. Two Dutch pilots, Jaap van Helsdingen and Guus Deibel, scored highest with the Buffalo with three victories each. Incidentally, just like the Finnish machines, the ML-KNIL Brewsters were not called 'Buffalo'. Basically this name was used for the British machines where they were known as the 'Buffalo I'. In the ML-KNIL it was simply called the 'Brewster jager' (=fighter) and the name Buffalo was adopted after the war!

Two examples show the nature of the aerial struggle. In a major fight on Java above Semplak on 19 February 1942, nine Dutch Brewster fighters intercepted a formation of five Japanese bombers with an escort of nineteen Ki-43's. Although the Brewster pilots claimed 11 Japanese aircraft no J.A.A.F planes were downed. The ML-KNIL suffered with the loss of four Brewsters; two Dutch pilots were killed in action. Two and a half weeks later, four Brewsters led by Captain Jaap van Helsdingen (who was not supposed

to be flying combat missions) with three other volunteers flew the legendary last ML-KNIL sortie on 7 March, one day before Dutch forces surrendered. As the flight encountered more than a dozen Japanese Ki-43's, van Helsdingen was last seen going down low over the jungle. His plane and his body were never recovered. Post-war he was posthumously awarded the Knight Officer medal of the Military Order of William.

After the surrender of the Netherlands East Indies on 8 March 1942, and the fall of Singapore to Japanese troops a small number of RAF and ML-KNIL Buffalos was captured by the Japanese. Several were fitted with Japanese markings and exhibited on a display of captured aircraft. When drawing conclusions, the Buffalo was not the most modern fighter the ML-KNIL could use against Japan but under the conditions of being outnumbered and trying to defend enormous areas the Buffalo did not score very bad. It definitely was not the failure as sometimes stated in post-war literature!

Japanese markings

A captured ML-KNIL Model 339 in Japanese markings. It was flown by Japanese pilots and exhibited together with other captured allied aircraft..

AS430

An ex-Belgian Model 339B in R.A.F. colours as AS430 in August 1940 at Boscombe Down It was used for test flights until it was damaged and lived further as instructional airframe 2859M until it was scrapped.

(Mark Nankivil collection)

R.A.F. service:

Great Britain received the largest number of Brewster fighters. Their first batch of 33 consisted of Model 339B aircraft originally intended for Belgium. They were delivered by ship and received the RAF serials AS410 to AS437, AX811 to AX820 and BB450. Although these serials imply 39 aircraft, they include the six stranded on Martinique for which the British acquired title. The Brewster was extensively tested at Boscombe Down; the ex-Belgian AS412 being the first. After an engine failure in July 1940 it was replaced by AS425. However, this aircraft was found to be structurally damaged and again it was replaced; this time by AS430. When this plane collided in October 1940 with a Hampden bomber during an air-to-air photographic session it had to be replaced again; AS426 was used for performance trials and AS410 for

Belgian in the U.K.

Buffalo I AS417 was an ex-Belgian machine. Most likely this photo was taken at Boscombe Down as a Ministry of Supply image. *(Mick Gladwin collection)*

No. 453 Squadron

This row of Buffalo I's was stationed at Singapore when this photo was taken and shows a line up of no.453 Squadron. Most pilots of this squadron were also Australians.
(Mark Nankivil collection)

armament evaluation. Three Brewsters were assigned for a short period to No. 71 Eagle Squadron until they were replaced by Hurricanes. A few other Brewsters were used at Fleet Air Arm training units like No. 759 Squadron. A handful were used on the aircraft carrier HMS Eagle and since they were not fitted with an arrester hook the pilots used their tail wheel to catch the arresting cable! Another twelve were assigned to No. 805 Squadron of the Fleet Air Arm and operated alongside Fairey Fulmar fighters. Three were used briefly over Crete during March 1941. When Crete was lost to the Germans, the remaining Brewsters of No. 805 Squadron were based at Dekheila near Alexandria in Egypt. They played hardly any essential role in the air war and one (AX813) was shot down by German fighters.

The British also paid for aircraft made to their specific order, the Model 339E. These followed later with the delivery of 120 with R.A.F. serials W8131 to W8250. A second order of 50 included AN168 to AN217. In the U.K. the Model 339E was named the 'Buffalo I'. In fact the name 'Buffalo' was only officially given to the R.A.F. machines; the name was not used in Finland and in the Netherlands East Indies were they were simply known as 'Brewster fighters'. The name 'Buffalo' was only used for these machines in post-war publications! For U.S. Forces (where the official F2A-1/2/3 designations were used), popular names were officially adopted in October 1941, and Navy and Marine personnel did use 'Buffalo' in wartime reports.

Japanese visit!

Brewster AN196 was one of the Buffalos from No. 243 Sq. captured quite intact by the Japanese together with some ML-KNIL B-339's. It is here inspected by a Japanese soldier.
(Mark Nankivil collection)

Of the 170 aircraft in the Model 339E order, the first three (W8131 to 33) were sent to Great Britain for testing. Buffalos W8132 and W8133 were tested at Boscombe Down before being relegated to instructional airframe use. In general the pilots flying the Buffalo were very positive on its flying characteristics. It was easy to fly, easy to land and highly manoeuvrable. However, its performance was poor when compared with other operational fighters and the Buffalo was regarded as unsuitable for the war over Europe.

Therefore, the British Buffalos found their way to South-East Asia, stationed at Singapore, Malaya and Burma. They were sent crated to Singapore or Rangoon by ship and assembled during the spring and summer of 1941. By December 1941, approximately 150 Buffalo Model 339E aircraft made up the bulk of the British fighter defence of Burma, Malaya and Singapore. They were assigned to Nos. 21, 243, 453 and 488 Squadrons and to 4 Photo Reconnaissance Unit on Singapore Island and Malaya and 67 Squadron in

Burma. These locally formed squadrons consisting of a mix of RAF, RAAF and RNZAF pilots. The squadrons operating from Singapore flew in cooperation with a small number of ML-KNIL Brewsters.

The RAF squadrons were newly formed and when Japan attacked most pilots had only very limited flight experience and, except for a few transferred from Europe, absolutely no battle experience. In fact a great number of them had just completed a few hours of solo flying on

Buffalo Mk.I 'AN185' was flown in Singapore by Australian pilots of no. 453 Sq.. Its regular pilot was Richard Douglas Vanderfield, who claimed five victories during the war with Japan.
(©Srecko Bradic)

the Wirraway (an Australian development of the Harvard trainer)! More serious was the fact the British Buffalo's were all fitted with the 1100 hp Cyclone engine. Additional equipment and armour had further increased their all-up weight and, just before the war began, the British pilots discovered that the ML-KNIL Brewster 339C had far superior performance compared to their 339E Buffalos!

When Japan attacked in December 1941 the Buffalo squadrons sustained heavy losses. In general they fought an uneven battle against overwhelming Japanese forces. The airfields were regularly attacked by Japanese bombers and many allied aircraft including Buffalos were destroyed on the ground. Still they fought on under worsening conditions with aircraft that were rapidly worn out under the tropical conditions. When Japan had captured Burma and the Malayan peninsula, the Buffalos remaining were barely in airworthy condition and when Japanese

forces captured the airfields they found several Buffalos. Some of the machines from Burma withdrew to India and served with 5 and 146 Squadrons, and eventually with the wartime Indian Air Force. In spite of their poor performances, the R.A.F. Buffalo I still produced four Commonwealth aces:
- Geoff Fisken
- Maurice Henry Holder
- Alfred Wattle Benjamin Clare
- Richard Douglas Vanderfield.

New Zealander Fisken, the top-scoring pilot with six victories as a Buffalo pilot, later flew P-40 Kittyhawks in the Royal New Zealand Air Force and claimed five more. With a total of 11 he became the highest-scoring Commonwealth pilot in the Pacific theatre.

Maurice Henry Holder claimed 5 victories while flying Buffalo's at R.A.F. 243 Sq.(2 and 3 shared destroyed, and 2 shared damaged). He returned to the UK, only to

be killed in a training accident on 16 July 1942. Clare and Vanderfield both claimed five aerial victories when flying at the Australian No. 453 Squadron.
During the whole Southeast Asia campaign, the British Buffalos lost about 40 aircraft in air combat (and more were destroyed on the ground during bombing raids). Against this, 80 aerial victories were claimed.

Dutchy in the U.S.A.

In-flight publicity shot of ML-KNIL Brewster B-339 flying over New Yersey with both Dutch military and U.S. civil markings.
(Mark Nankivil collection)

RAAF/USAAF use:

As we have seen earlier the last batch of 20 Model 339-23's ordered by ML-KNIL was diverted to Australia. They carried ML-KNIL registrations B-3167 to B-3186. They resembled denavalized F2A-3's but were fitted with 1000 hp Cyclone export version engines.

In fact it was an unlucky combination of a high all-up weight and the lower rated engine. At least they had an excellent range, some pilot armour and self-sealing fuel tanks but it is retrospectively maybe better they were never employed by ML-KNIL against the Japanese. In fact they were unsuitable to be used as a fighter and they were the slowest Buffalo's built with a maximum speed at sea level of only 424 km/h. They arrived crated in Australia by Dutch or Netherlands East Indies ships and were assembled at either the International Harvester Works at Geelong or by Ansett Airlines at Essendon over the period March-April 1942. Initially they were taken on by the U.S.A.A.F.

Dutchman in Australia

B-3119 was the first ML-KNIL B-339D fitted with the 1200 hp Cyclone engine. It was not delivered to ML-KNIL since it was damaged in the U.S.A. in a landing incident. When repaired it was diverted to Australia and finally fitted with U.S.A.A.F. markings although it retained its ML-KNIL colour scheme and its original registration '3119'.

(Jim Maas collection)

No. '313' must have been the last airworthy U.S.A.A.F. Buffalo! It is shown here in a highly polished scheme with late style U.S. national markings. It was flown by a U.S. squadron commander from Essendon near Melbourne.
(©Srecko Bradic)

The same 'B-3119' as flown in Australia with markings of the U.S.A.A.F. and with U.S. ARMY under its wing.

(©Srecko Bradic)

The Last
The following details on the identity of the Model 339-23's could be traced:

ML-KNIL	R.A.A.F. No.	U.S.A.A.F. No.	Remarks
B-3167	A51-13	310	
B-3168	A51-12	309	
B-3169	?	?	
B-3170	?	?	
B-3171	A51-1	n/a	n/a: not applicable
B-3172	A51-7	301	
B-3173	?	?	
B-3174	A51-14	311	remains at Aviodrome
B-3175	A51-2	n/a	
B-3176	A51-9	305	
B-3177	A51-3	n/a	
B-3178	A51-10	307	remains at Aviodrome
B-3179	A51-11	308	
B-3180	A51-16	303	
B-3181	A51-8	302	
B-3182	A51-17	306	
B-3183	A51-15	312	
B-3184	A51-4	n/a	
B-3185	A51-6	n/a	
B-3186	A51-5	n/a	

5th Air Force based in Australia. Of the twenty, seventeen were passed to the Royal Australian Air Force and assigned, at different times, to R.A.A.F. 24, 25 and 85 Squadrons and to 1 PRU. They received R.A.A.F. registrations: A51-1 to A51-17, although these numbers did not follow their ML-KNIL serial sequence, nor the three digit tail numbers (from 301 to 313) given to many by the 5th Air Force. Also a single Model 339D (ML-KNIL registration B-3119), left behind in the U.S.A. after it was damaged in a crash, was finally shipped to Australia after it had been repaired. It retained its original registration 3119 (without the prefix 'B) and went also to the 5th AF. In its original ML-KNIL camouflage colours, but with U.S. markings and with 'U.S. ARMY' under its wings it was used most likely as a liaison aircraft. Its fate remains unknown but must likely it was left behind when no longer serviceable and scrapped. One U.S.A.A.F. Buffalo, with registration no. 313, was used as a personal aircraft by a local Unit commander at Essendon near Melbourne in highly polished bare metal colours and used as late as 1944. When no longer serviceable it was abandoned in a corner of the airfield and later scrapped. None of the Australian/U.S.A.A.F. Model 339-23's encountered any enemy aircraft during their flying careers.

Australian/U.S.A.A.F.

Rare image of Model 339-23 with the registration nos. 312/A51-16 in U.S.A.A.F. markings. Photo was taken by Ed Jendzeizyk when he was with the U.S. Army in Australia.

(Jim Maas Collection)

Großbritannien und USA.

Brewster „Buffalo" (F 2 A-2) [1]

Jagdeinsitzer
(auch für Trägereinsatz)

[1] USA.-Werksbezeichnung: **Brewster 439.**

Recognition

Second page of the Brewster Buffalo chapter from the German aircraft recognition manual 'Die Kriegsflugzeuge der Feindmächte'

Museum wreck

Wreck of Lauri Pekuri's BW-372 as currently displayed at the Aviation Museum of Central Finland.

Survivors:

Only one complete original Brewster fighter, with Finnish registration *BW-372*, still exists. It was recovered from a lake in Russia and was found to be quite complete. It is on display as it was recovered at the Aviation Museum in Central Finland.

Two Model 339-23 wrecks from crashed machines (*A51-10* and *14*) were partly recovered in Australia and finally shipped to the Netherlands for display at the Aviodrome museum at Lelystad airport. They are nothing more than incomplete remains of what was found and at present they are stored. The state of this shipment of scrap (and it is nothing more than this) makes a future restoration as a partial static exhibit quite impossible.

In July 2008, a wooden replica Model 339C was completed by the Cradle of Aviation Museum in Long Island, New York. The aircraft carries the markings of an ML-KNIL fighter flown by Lt. Gerard Bruggink (two kills). It was built for the Militaire Luchtvaart Museum (Military Aviation Museum) at Soesterberg, the Netherlands. The Cradle of Aviation Museum also made a replica F2A-2, carrying the markings of unit '*201-S-13*' from VS-201, aboard USS *Long Island*.

Replica of Dutch Buffalo

A replica Model 339C in storage at the Soesterberg airbase facilities for future exposition in the new Dutch Armed Forces Museum.

Details

Some detail shots of the Model 339C replica.

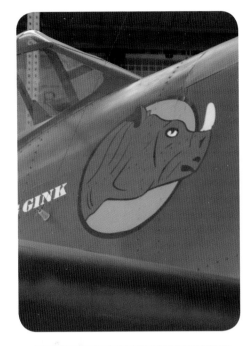

Technical details:

Specifications ot the various versions:

Powerplant: Wright R-1820 Cyclone nine-cylinder single-row air-cooled radial engine

XF2A-1	R-1820-G22	950 hp
F2A-1	R-1820-34	950 hp
XF2A-2	R-1820-40	1200 hp
F2A-2	R-1820-40	1200 hp
F2A-3	R-1820-40	1200 hp
Model 239	R-1820-G5	1000 hp
Model 339B	R-1820-G105	1100 hp
Model 339C	R-1820-G105A	1100 hp
Model 339D	R-1820-G205	1200 hp
Model 339E	R-1820-G105	1100 hp
Model 339-23	R-1820-G5E	1000 hp

Dimensions:

Length:

XF2A-1	7.925 m
F2A-1	7.938 m
XF2A-2	7.807 m
F2A-2	7.807 m
F2A-3	8.050 m
B-239	7.938 m
B-339	7.987 m
B-439	8.220 m

Wingspan: 10.67 m for all versions

Height: 3.66 m for all versions except XF2A-1 (2.70 m) and XF2A-2 (3.50 m)

Wing area: 19.4 m² for all versions

Weights:

Empty:

XF2A-1	1688 kg
F2A-1	1728 kg
XF2A-2	1883 kg
F2A-2	1883 kg
F2A-3	2162 kg
B-239	1702 kg
B-339B	1823 kg
B-339E	1936 kg
B-439	2150 kg

Loaded:

XF2A-1	2268 kg
F2A-1	2293 kg
XF2A-2	2354 kg
F2A-2	2455 kg
F2A-3	2957 kg
B-239	2278 kg
B-339B	2466 kg
B-339C	2620 kg
B-339D	2684 kg
B-339E	2772 kg
B-339-23	2950 kg

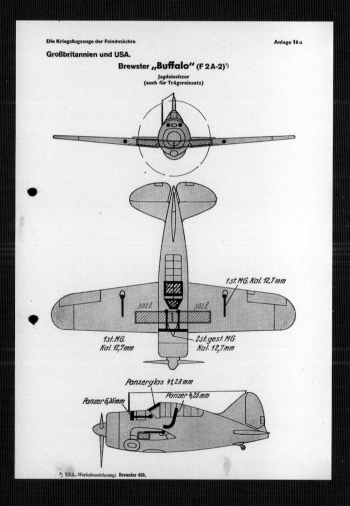

Although it was never used in Europe except by the Finns, the Buffalo was included as a chapter in the aircraft recognition manual of the German Kriegsmarine.

Performances:

Max. speed:		km/h at 4500 m	km/h at sea level
	XF2A-1	489	475
	F2A-1	500	431
	XF2A-2	535	475
	F2A-2	554	466
	F2A-3	517	457
	B-239	483	465
	B-339B	527	471
	B-339C	528	489
	B-339D	545	494
	B-339-23	494	424

Range:		
	XF2A-1	1500 km
	F2A-1	1600 km
	XF2A-2	1500 km
	F2A-2	1500 km
	F2A-3	3000 km
	B-239	1600 km
	B-339	1650-2000 km
	B-339-23	

Service ceiling:	XF2A-1	9400 m
	F2A-1	10,180 m
	XF2A-2	10,120 m
	F2A-2	10,330 m
	F2A-3	9100 m
	B-239	9900 m
	B-339	10,200 m
	B-339-23	approx. 9000 m

Crew: One, pilot

Armament:	machine guns
XF2A-1	unarmed
F2A-1	1x 12.7 mm and 1x 7.62 in the fuselage + 2x 12.7 mm in the wings
XF2A-2	2x 12.7 mm
F2A-2	4x 12.7 mm
F2A-3	4x 12.7 mm
Model 239	3x 12.7 mm + 1x 7.62 mm (later 4 x 12.7 mm)
Model 339B	2x 12.7 mm in the fuselage + 2x 7.62 in the wings
Model 339C/D	2x 7.62 mm in the fuselage, 2x 12.7 mm in the wings
Model 339E	4x 12.7 mm (although at some R.A.F. machines the wing 12.7 mm were replaced by 2 x 7.7 mm (.303 cal)
Model 339-23	4x 12.7 mm (some were unarmed)

Group editor	Corrections
Edwin Hoogschagen	Jim Maas
Editors	**Graphic design**
Nico Braas	Jantinus Mulder
Srecko Bradic	
	Publisher
Author	Lanasta
Nico Braas	

First print, January 2013
ISBN 978-90-8616-164-5
NUR 465

Contact Warplane:
Slenerbrink 206, 7812 HJ Emmen
The Netherlands
Tel. 0031 (0)591 618 747
info@lanasta.eu

Violearo

© Copyright 2013 Lanasta, Emmen

www.lanasta.com

All rights reserved.
All correspondence regarding copyrights, translation or any other matter can be directed to: Lanasta, Slenerbrink 206, 7812 HJ Emmen, The Netherlands.

References

-Anonymous, Brewster's benighted Buffalo, Air Enthusiast Quarterly No. 1 p. 66-83
-Gerard Casius, België en Ned. Indië kochten Brewster Buffalo, Aviahobby July 1980 p. 124-133
-Gerard Casius, Brewster Buffalo, Aviahobby Sep.1980 p. 176-179
-Jean-Louis Couston, Le Brewster Buffalo, Lela Presse-France (2004)
-C.C. Küpfer, Onze vliegers in Indië, Boom-Ruygrok N.V. (1946)
-Malcolm A. LeCompte, Naval air supremacy and the development of the Brewster Buffalo, AAHS Journal Vol.41 no. 4 (1996)
-Jim Maas, F2A Buffalo in action, Squadron Signal Publications 1987
-Tim Mason, The Secret Years- Flight testing at Boscombe Down 1939-1945, Crécy Publishing Ltd UK (2010)
-Gordon Swanborough & Peter Bowers, United States Navy aircraft since 1911, Putnam UK (1976)

Illustrations

All photographs: author's collection unless stated otherwise

Credits:

We specially thank Jim Maas for correcting and proof-reading of the manuscript.
We also thank the following persons for providing photographs:
Daniel Brackx, Gerard Casius, Frits Gerdessen, Edwin Hoogschagen, Loet Kuipers, Jim Maas, Mark Nankivil, Kari Stenman and the Finnish Air Force Museum.